Resourceful Rain Forest

BY
PAT AND BARBARA WARD

Printing No. CD–1305

Mark Twain Media, Inc., Publishers
Distributed by Carson-Dellosa Publishing Company, Inc.

Table of Contents

Prologue

"Well, Dear," Henry said, "it's such a nice day; let's have coffee on the patio." Henry sat down on his new teakwood bench, took a big drink of orange juice, and opened the daily newspaper. "Good grief," he growled, "I'm tired of all these stories about the importance of the rain forest! Who cares? The rain forest is thousands of miles away."

Henry's wife, Marge, walked in with breakfast: a banana, a grapefruit, and eggs with black pepper. She set the food down on the rattan table. "What, no coffee?"

"No, Dear, the doctor said tea was better for you. This tea is vanilla-flavored. You'll like it. And don't forget your high blood pressure medicine."

"Okay, okay," Henry said from behind his paper. "Hey, look at this, Marge. This picture of the rain forest has plants that look just like that potted fig tree over there. And by the way, I noticed some white flies on it the other day. Do we have any Rotenone insecticide left?"

As Marge left to go check, Henry staked his favorite orchid with a piece of bamboo. He munched on a handful of chocolate-covered macadamia nuts. *So what's all this rain forest talk*, Henry mused. He took a deep breath of clean, fresh air. What a nice day.

What Is a Tropical Rain Forest?

When we are studying a new subject, the first thing we should do is figure out exactly what we are talking about. Let's try to figure out what we mean when we are talking about a tropical rain forest. First of all, it is a forest, which means it has a lot of trees. It is characterized by warm temperatures all year. A tropical rain forest also has more than 60 inches of rain per year! It also has very high humidity.

Now that we have an idea of what we are talking about, let's try to figure out where we can find them. Tropical rain forests are found in a band around the earth's equator. This band is bordered on the north by the **Tropic of Cancer** and on the south by the **Tropic of Capricorn.**

Since the tropical rain forests are near the equator, they get about 12 hours of sunlight every day. The amount of sunlight does not change from month to month during the year. The Sun's rays hit the earth at right angles at the equator, so this area gets very intense and warm sunlight. The temperature does not usually change very much. It ranges from 70 to 85 degrees Fahrenheit most of the time.

As we said, rain is common in the tropical rain forests. Accumulations of 200 inches per year are not uncommon, and the record yearly rainfall stands at over 400 inches! All of the heat and rain causes very humid conditions. The humidity is usually 70 percent during the heat of day and 95 percent at night.

Tropical rain forests cover only about seven percent of Earth's land surface. However, they are home to about 50 percent of Earth's species of plants and animals. The plants of the tropical rain forest release huge amounts of water vapor into the air. They affect the weather over a large part of the globe.

The word *jungle* is often used to describe the tropical rain forest; however, the rain forest is not always a dense, impenetrable mass of vegetation. The first view that explorers and settlers had of the tropical rain forest was often from boats as they traveled down rivers. What they saw along the rivers really was a jungle. The rain forest takes advantage of open areas that provide light gaps, such as are often found along rivers and streams. Seedling trees, ferns, vines, and shrubs grow in great numbers. From the water, these plants screen the interior of the rain forest. Nevertheless, the interior itself is actually much more open. There is not enough light for thick plant growth along the forest floor.

So, now you should have a better idea of what we mean when we talk about a tropical rain forest. How many of you have ever lived in a tropical rain forest or even visited one? We will learn more about these wonderful ecosystems and how they can affect us.

Name ______________________________ Date ______________

For the Student:

1. What are three characteristics of a tropical rain forest?

2. Do rain forests cover a large percentage of Earth's land surface? Why?

3. Why are the Sun's rays more intense at the equator than they are in North America?

4. There are rain forests in the states of Oregon and Washington. Are they tropical rain forests? Why?

5. What aspects of the tropical rain forest could affect global weather patterns?

Matching:

_____	6. Southern border of the tropics	A. Forest
_____	7. Northern border of the tropics	B. Humid
_____	8. Dense, impenetrable mass of vegetation	C. Jungle
_____	9. An area with a lot of trees	D. Tropic of Cancer
_____	10. Heat and rain cause these types of conditions	E. Tropic of Capricorn

Off to the Tropical Rain Forest

Millions of years ago, tropical rain forests covered much of the earth's surface. Through the years, the climate on Earth changed. These climatic changes caused the tropical rain forests to shrink.

Three hundred years ago, a band of tropical rain forests stretched around the earth in an almost continuous green carpet. It grew in parts of South America, Africa, Southeast Asia, Indonesia, and Australia. Before people began to cut and burn the trees, these forests may have covered more than 20 percent of the earth's surface.

Today, the tropical rain forests have been severely reduced. They are now a series of green "islands," although some of them are quite large. About 30 countries have at least some tropical rain forest. More than half of all of the rain forests remaining are located in three countries: Brazil in South America, Zaire in Africa, and in parts of Indonesia. The largest single area of remaining tropical rain forest is found in the basin of the mighty Amazon River in Brazil. For those of us living in the United States, the closest forests that might be classified as tropical rain forests would be those found in southern Mexico, in Central America, and on some of the Caribbean islands.

There are two other types of forests that have some characteristics similar to those of the tropical rain forests. One of them is called the **tropical seasonal forest**. This type of forest has a very definite wet season and a definite dry season. Remember that true tropical rain forests have lots of rain all year! The other type of forest is called the **tropical cloud forest**. This kind of forest is found in mountain areas. The moisture for this type of forest comes from mists and clouds.

A tropical cloud forest

Name ______________________________ Date ______________

For the Student:

1. What three countries have the most tropical rain forests?

____

2. What are two things that have caused a reduction in the tropical rain forests?

____

____

3. Is there a tropical rain forest in the USA? Why?

____

____

4. What are the seasons of a tropical seasonal forest?

____

____

5. Where does the tropical cloud forest get its moisture?

____

____

Locate and color in the areas of tropical rain forest on the world map below.

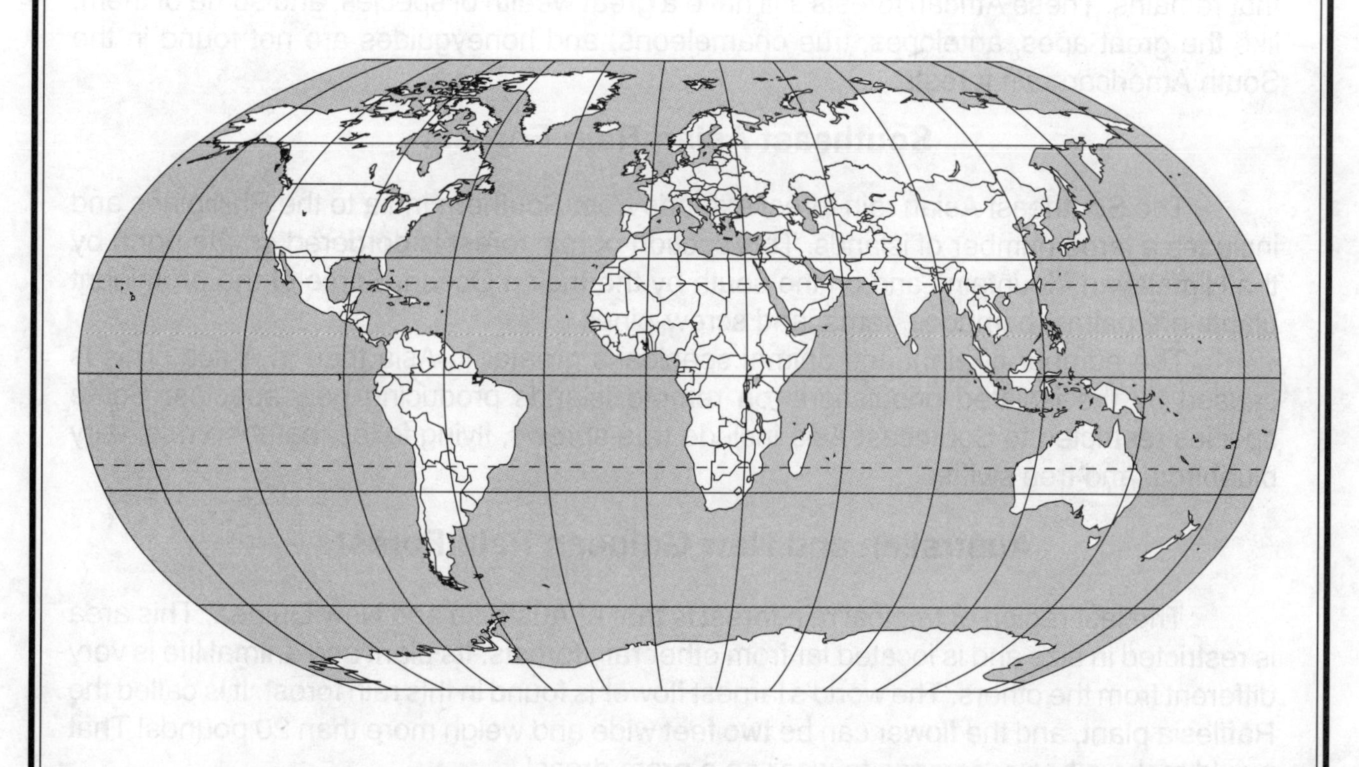

Regions of Tropical Rain Forest

The South American Rain Forest

The South American rain forest is located in the northern half of South America. It begins east of the Andes Mountains and extends north to Central America. Parts of Central America, through the Gulf of Mexico, and some islands of the West Indies also have tropical rain forests.

Tropical rain forests look similar wherever they are found. There are, however, distinct differences in plant and animal life. For example, Arrow-poison frogs, bromeliads, cacti, hummingbirds, tucans, and iguanas are common in some areas of the South American tropical rain forest, but they are not found in Africa or Asia.

African Rain Forests

West Africa is the smallest of the three major rain forest areas. Some patches of rain forest occur across eastern Africa all the way to the Indian Ocean. We think of rain forests as being full of plants and animals. The African rain forest, however, has a lower overall density of species. Some plants that are very common in American and Asian rain forests are surprisingly scarce in Africa. These include palms, orchids, and epiphytes. The number of bird species at one site in the African rain forest is about 200 species. A bird list from a site in the South American rain forest may number over 500 species. This lower bird and plant density may be related to the smaller size of the African rain forest.

It also lacks major barriers such as large ruins and mountains. Breaks like that might tend to separate populations and increase the number of species. The African rain forests do not have as many breaks, so the species compete with each other in the amount of forest that remains. These African forests still have a great wealth of species, and some of them, like the great apes, antelopes, true chameleons, and honeyguides are not found in the South American rain forests.

Southeast Asian Rain Forests

The Southeast Asian rain forest extends from Southern India to the Phillipines and includes a large number of islands. This section of rain forest is bordered on the north by the Himalayan Mountains and on the south by the Indian Ocean. Some of the prominent plants are palms, bamboos, ferns, and screw-pines.

The number of plant and animal species is greater in Asia than in Africa. This is caused by the isolated populations on remote islands producing new species. Some species restricted to Southeast Asia include true shrews, flying foxes, bamboo rats, fairy bluebirds, and tree swifts.

Australian and New Guinean Rain Forests

The last region of tropical rain forest is that of Australia and New Guinea. This area is restricted in size and is located far from other rain forests. Its plant and animal life is very different from the others. The world's largest flower is found in this rain forest. It is called the Rafflesia plant, and the flower can be two feet wide and weigh more than 20 pounds! That would make a heavy corsage to wear on a prom dress!

Name ______________________________ Date ____________________

For the Student:

1. Which area of tropical rain forest has the greatest diversity of life and why?

____

____

2. Do the Indians of the Amazon have to keep an eye out for great apes and tigers? Why?

____

____

3. What are some natural barriers that can increase the number of species in an area?

____

____

____

Match the plant or animal with the rain forest from which it comes.

_____ 4. Bamboo rats
_____ 5. Great apes
_____ 6. True shrews
_____ 7. Rafflesia
_____ 8. Tucans
_____ 9. True chameleons
_____ 10. Arrow-poison frogs
_____ 11. Fairy bluebirds
_____ 12. Honeyguides
_____ 13. Iguanas
_____ 14. Flying foxes
_____ 15. Hummingbirds

A. South American Rain Forest
B. African Rain Forest
C. Southeast Asian Rain Forest
D. Australian and New Guinean Rain Forest

Activities:

1. Pick one of the plants or animals mentioned in this chapter and write a short report on it.
2. Find the definition for the following:
   - A. Bromeliad
   - B. Epiphyte
   - C. Cacti
   - D. Honeyguide

Name ______________________ Date ______________

Activity: Travel in the Rain Forest

You have been invited to travel to a school to talk about life in the United States. The school is located in Ecuador, on the Napo River. The Napo River is a tributary of the Amazon River. Some of the students are Indians, and some of the children's parents are gold prospectors. The prospectors traveled up the Amazon and Napo Rivers from Brazil. You will be sent an airline ticket to Lago Augrio, a bus ticket to the Napo River landing, and the name of a dugout canoe boatman. You will be gone one week. You need to pack lightly.

Give each item listed a number:
1 = essential
2 = nice to have
3 = do not need

_____ 1. passport
_____ 2. gloves
_____ 3. long-sleeved shirt
_____ 4. anti-malaria medicine
_____ 5. sandals
_____ 6. bug spray
_____ 7. insulated boots
_____ 8. 2 pairs of socks, underpants
_____ 9. Hard Rock Cafe t-shirt
_____ 10. tennis shoes
_____ 11. waterproof backpack
_____ 12. swimsuit
_____ 13. Pepto Bismol
_____ 14. wash towels
_____ 15. aspirin
_____ 16. plastic bags
_____ 17. binoculars
_____ 18. 8 pairs of socks, underpants
_____ 19. Spanish dictionary
_____ 20. lined jacket
_____ 21. shorts
_____ 22. suntan lotion
_____ 23. toothbrush, toothpaste, deodorant
_____ 24. $10–$100 traveler's checks
_____ 25. $200 in small bills
_____ 26. video tapes of your school
_____ 27. water purification pills
_____ 28. Portuguese dictionary
_____ 29. First-aid kit
_____ 30. mosquito netting, sleeping bag
_____ 31. rain coat
_____ 32. snake bite kit
_____ 33. umbrella
_____ 34. Kleenex

Life Layers

Life in the tropical rain forest is arranged in a series of levels or layers. Each kind of plant and animal is adapted to live at a certain level. The upper layer, for example, receives the most sunlight. This is the area where we would expect plants that need the most light. The forest floor receives the least amount of sunlight. It would be natural for us to think that the plants that grow there must be tolerant of shade. Temperature also changes from the upper to the lower levels. Temperatures can even change within a level. Temperatures at the upper level fluctuate the most. The forces of the rain and wind are also more severe at the upper levels. The most stable environments are at the lower levels.

It rains often in the tropical rain forest. (I guess that is why they call it the rain forest!) The humidity from the frequent rains is trapped in the shady lower levels. That leaves the highest levels to be the driest. Scientists call the highest level the **emergent layer**. There are very tall trees that grow above the main canopy of trees. Most of these trees will be 100–150 feet tall, but some may grow to be 250 feet in height. These tall trees are most exposed to the problems caused by too much heat, heavy rainfall, and strong winds. They also receive the most sun to carry on their growth processes through photosynthesis. There are usually only one or two of these forest giants per acre. The base of some of these emergent trees may have very thick, wide, vine-like structures. These structures, called **buttresses**, may help support these wind-exposed trees far above the forest floor.

The **canopy layer**, usually 65–100 feet above the forest floor, is the continuous mass of green that forms the top of the forest. These trees receive the same light intensity, wind forces, amount of rain, and high temperatures. Most of these trees have to start life in the shade on the forest floor. They grow swiftly to the canopy, where more light is available. The limbs of these trees support **epiphytes**, which are plants that grow on other plants. The limbs also provide a structure for clingy vines. The leaves of the canopy trees are often pointed. This shape forms a **drip tip** that helps keep the leaves dry and safe from molds and lichens.

The plants of the **understory**, or **shrub layer**, have to adapt to less sunlight, but they also receive less heat, wind, and rain. Many of these plants have extra large leaves to absorb more of the filtered sunlight that finally reaches the lower layers. Understory plants are usually less than 15 feet tall.

The lowest level is called the **forest floor**. Plants that depend on the sun to provide food for survival have a rough time there. Instead, ferns, herbs, and mosses that flourish on the trapped humidity can be found. Also, plants such as fungi, which can live on fallen leaves and other dead materials, survive in the dim light of the forest floor. Contrary to popular belief, the floor of the tropical rain forest is often quite clear and easy to walk through.

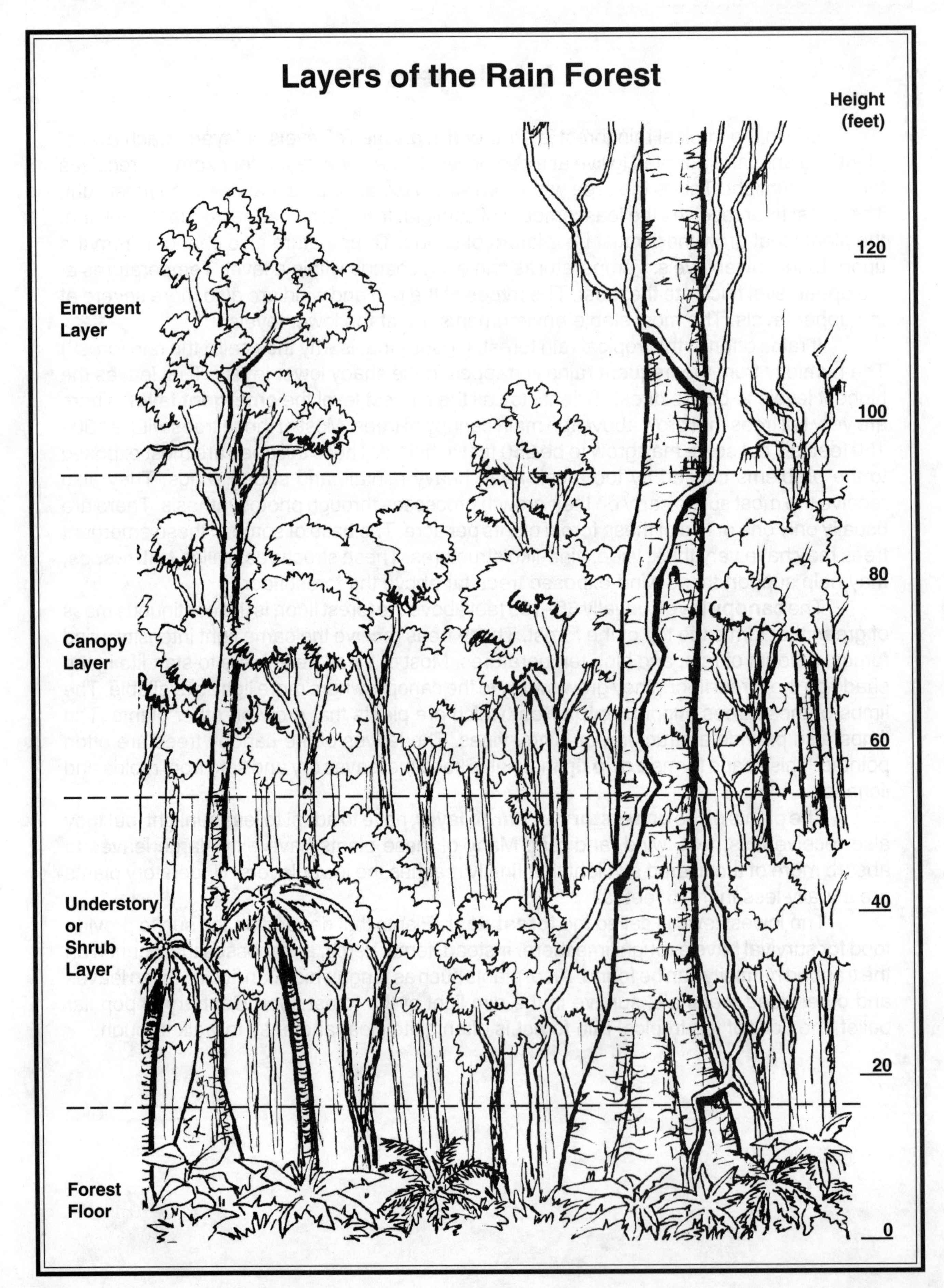
Layers of the Rain Forest
Height (feet)
120
100
80
60
40
20
0
Emergent Layer
Canopy Layer
Understory or Shrub Layer
Forest Floor

Name ______________________________ Date ____________________

For the Student:

1. What are the four levels of the tropical rain forest?

2. Which layer receives the most sunlight? Which receives the least amount of sunlight?

3. Why is vegetation sparse on the forest floor?

4. Why would epiphytes grow on the branches of trees instead of on the forest floor?

5. What is a "drip tip," and why is it important?

6. What are buttresses and what do they do?

7. Which level of the rain forest has the most stable conditions?

8. How many emergent trees are there per acre?

9. What types of plants do well on the forest floor?

10. Why do plants in the understory have extra large leaves?

Biodiversity

The professor asks: "How many different types of living organisms are found on your school grounds?"

"That's easy," you say. "There are some people, a dog, a couple of trees, some grass, and, oh yeah, a bird."

"Well," the professor says, "what about insects and other invertebrates and microorganisms living in the soil? What about the different types of grass? Are there any other species of plants? What kinds of trees are there? Are there lichens on the bark of the trees? What species of birds are there in the trees? You see, this is a rather complex question."

Think of how many different living organisms there are in a habitat such as a prairie. The number of living organisms in a habitat or a geographic area is known as its **biodiversity**. The earth's biodiversity is thought to be somewhere between 10 and 30 million species. Scientists are not sure of the number because new species are being discovered every year. Some groups of animals are pretty well-studied. There are about 4,000 species of mammals, and it is very unusual for a new one to be discovered. There are 9,000 species of birds, and there are a small number of new species discovered each year. As animals get smaller and are harder to see and to study, the number of possible new species increases.

Thousands, maybe millions, of insects are yet to be discovered, and scientists are just starting to study deep sea creatures. The world of microorganisms is also huge, and undescribed species abound.

Your school grounds have complex biodiversity. The biodiversity of a state park or a wildlife refuge is even more complex. We usually expect natural areas to be much more diverse than parking lots, city yards, or agricultural fields. The area of the world with the greatest biodiversity is the tropical rain forest.

The biodiversity of an area is important to the natural functioning of the area's ecosystem. The biodiversity can tell us a lot about the health of the environment. For example, an oil spill off the West Coast might kill off sea otters in a tidal lagoon. Sea otters eat sea urchins. The sea urchin population would increase in the tidal lagoon without sea otters preying on them. Sea urchins feed on seaweed or kelp. The sea urchins would graze down the seaweed or kelp beds in the tidal lagoon. Then, without this protective habitat, fish and invertebrates would disappear from the lagoon. The biodiversity would decrease. This would be a clear signal that the tidal lagoon was not a healthy ecosystem.

Protecting the world's remaining wildlands helps protect biodiversity. The greatest variety of species can be saved by protecting the tropical rain forest.

Name _______________________________ Date _______________

For the Student:

1. What is biodiversity?

2. Are there more species of birds or mammals in the world?

3. If you wanted to find a new species of animal, what would you look for, and where would you look?

4. Does a ten-acre field of corn or a ten-acre marshy wetland have a greater biodiversity?

5. What are ten species of living organisms that inhabit your school yard?

6. What area has the greatest biodiversity?

7. What happens when one species in an environment is killed off?

8. Why do scientists think there are so many species of insects, deep sea creatures, and microorganisms yet to be discovered?

Plants of the Tropical Rain Forest

When we plant a vegetable or flower garden, we select an area with lots of sunlight (for warmth), good soil, and water nearby. The tropical rain forest has all of these ingredients: **constant warmth**, **sunlight**, and **rainfall**. These conditions create a huge, complex garden with a great diversity of plants. The small country of Ecuador, for example, has a tropical rain forest among its diverse habitats. It has a plant list of more than 20,000 species. That's more kinds of plants than can be found in all of Europe! The Asian country of Malaysia has 2,500 species of trees. That's more than three times the number of tree species growing in the United States. One reason for the great diversity of plant life is that the rain forest has different layers of vegetation. Each layer has different kinds of habitats. Many rain forest plants have a particular place where they live. That place meets the plants' needs for survival. This area is called their **niche**.

Some plants have developed special adaptations to survive in the rain forest. Some types of small trees, ferns, orchids, mosses, and even cacti have developed the ability to live high above the ground. We call these plants **epiphytes**. These plants use their ability to grow above ground to get closer to the sun and away from the shade of the forest floor. These plants will grow on the branches of trees and on other exposed surfaces. Sometimes they will even grow on telephone wires!

So how do epiphytes survive since their roots are not anchored in the soil? Some epiphytes are able to store water in their tissues. Other epiphytes collect rain water in their cup-like leaves. Still others have aerial roots. The aerial roots absorb water from the moisture-laden air of the tropical rain forest.

How can an epiphyte get started growing 100 feet above the ground on the branch of a tree? Many epiphytes produce a fruit that birds just love to eat. The birds digest the fruit, but they excrete the seeds along with their other wastes. The epiphyte seeds are deposited wherever the bird drops them. Sometimes that happens to be on the branches of tall trees or even on telephone wires. Some big, strong tropical trees may have hundreds of epiphytes growing on a single branch. The weight of the epiphytes can actually break the branches off of the trees. Some trees have developed an extra layer of bark that can slough off, dropping the extra weight of the epiphytes to the forest floor below.

Other tropical rain forest plants must be rooted in the soil so they start their lives on the shady forest floor. Trees, some of which will become the giant emergent trees of the rain forest, start from seeds dropped to the forest floor. With the lack of sunlight, these trees grow very slowly. When a **light gap** occurs, such as when an old tree dies or a storm blows a tree down, these patiently waiting seedlings jump to life and grow quickly to fill the gap.

Other types of sun-loving trees have another method of using light gaps. The seeds of these trees lie dormant for years, waiting for a light gap to occur. When a gap opens up, the seeds germinate quickly and then grow rapidly. The winners of this growth race are the trees that reach the canopy first and take their places in the sunny canopy of the rain forest.

Another adaptation is found in the climbing vines. These plants must root in the soil of the forest floor. They can grow very rapidly, because they do not have to produce woody cells to support themselves. Instead, they cling to the bark of another plant, called the **host plant**. They grow very quickly, moving up toward the sunshine. Sometimes these vines will even cover the host plant and completely shade it.

Name ______________________________ Date ______________

For the Student:

Read each sentence below. Decide if the sentence is true or false. If a sentence is false, circle the F in front of the sentence. If a sentence is true, circle the T in front of the sentence.

T F 1. All plants must have roots in the soil to survive.

T F 2. Malaysia has more species of trees than the United States.

T F 3. Epiphytes live without water.

T F 4. Vines wait for a light gap to grow.

T F 5. Epiphyte seeds are spread by birds.

T F 6. All tree seeds germinate the first year.

T F 7. Some ferns and mosses are epiphytes.

T F 8. Vines support themselves by clinging to their hosts.

T F 9. A light gap can occur in the rain forest floor when a large tree dies.

T F 10. Ecuador has a plant list of about 2,000 species.

11. What is a niche?

____

____

12. How do some trees get rid of epiphytes?

____

____

13. What are aerial roots?

____

____

14. What occurs when a large tree dies and leaves a light gap?

____

____

____

____

Plant Defenses

There is plenty of rain and warmth for plant growth in the tropical rain forest, but this does not mean that life is easy there. To begin with, there is a lot of competition for sunlight. Next, plants that do manage to get the sunlight they need must be able to survive a multitude of leaf-eating and wood-munching animals.

Plants have developed a number of defenses to protect themselves. Some plants have developed leaves that have a horrible taste. Some of them have even developed poisonous chemicals in their leaves. Of course, this discourages leaf-eating animals. Other plants have bristly leaves, and some of them have thorns. As you might imagine, it would not be much fun to bite down on a mouthful of bristles or thorns.

One kind of vine which is eaten by the caterpillar of a certain butterfly grows protrusions on its leaves that look like butterfly eggs. An egg-laying butterfly will pass that vine up because it sees the fake eggs and thinks the plant food has already been taken.

Some plants actually have a latex, or rubber-like, substance in their leaves. Insects that try to feed on these leaves get a gummed-up mouth.

Moist conditions in the tropical rain forest allow for mold and other fungal growths. Most tropical rain forest plants have a point to their leaves. This is called the **drip tip**. This tip helps the leaf shed water and stay drier. It is harder for the molds and fungi to grow on the dry leaves.

Many plants of the tropical rain forest have very smooth or waxy leaves. This also helps prevent fungal diseases. The molds and other fungi have a hard time “holding on” to the smooth, waxy leaf surfaces.

Some plants produce a sugary substance that attracts ants. The ants are very protective of their plants. They will bite anything that tries to move into their area. The presence of the ants may help keep other insects, and even mammals, from feeding on the plants. The ants do not harm the plants, because they have their sugar to keep them happy.

Several common house plants such as philodendrons and fiddle leaf ferns have very large leaves. They make good house plants because they flourish under low light conditions. Large leaves also allow plants to survive in low light on the forest floor.

Bromeliads show another example of tropical rain forest plant defenses. Bromeliads are **epiphytes**, which means they do not have roots anchored in soil. They get their moisture and nutrients in an interesting way. Their leaves form a cup which fills with water during the tropical downpours. This water is available to the plant, but it is also used by many other animals. Insects, such as dragonflies, even lay eggs in the cup, and their young develop there. Some frogs live in these cups of water, and their tadpoles also develop there. All the activity associated with the insects and other animals leaves behind a residue in the leaf cup. That residue provides nutrients that the plant needs to survive.

Name ______________________________ Date ____________________

For the Student:

1. Why do leaves of some tropical rain forest plants have a point?

2. Why do some plants have poisonous chemicals in their leaves?

3. Why do some plants produce very large leaves?

4. What are two functions of a bromeliad's cup?

5. What do the following words mean? Use a dictionary to help you write a good definition for each word:

a) residue ____________________

b) nutrient ____________________

c) fungi ____________________

6. How do ants help protect some plants?

7. What happens to insects that try to eat leaves containing a latex substance?

8. Why do some plants have very smooth or waxy leaves?

Animals of the Tropical Rain Forest

Animals are often very hard to observe in the tropical rain forest. One reason they are hard to see is because visibility is often reduced by the low light of the forest floor. Remember that it is hard for sunshine to reach the forest floor.

Another reason that the animals are hard to see is the profusion of tree trunks, branches, and foliage. It is hard to find anything when you have brown trunks and branches in your way, as well as lots of green leaves and different colored flowers.

Yet another reason is because many animals in the tropical rain forest are also well-**camouflaged**. Think about a jaguar. It has mottled spots. They help the jaguar blend in with the diffused and spotty light that makes it to the forest floor. The Three-toed Sloth actually grows a mat of green algae on its fur. This algae helps hide the slow-moving animal for days.

Finally, another reason the animals are hard to see is because there are lots of different kinds of animals. However, their actual populations are often low, and so individual animals are scattered over a large area.

Some animals, such as the Collared and White-lipped Peccarys, travel in small herds or family groups. You will not see the great herds of animals that are often associated with grasslands or prairie areas.

Some tropical rain forest species are so secretive and rare that scientists are still finding new ones every year. Observing these animals in their natural environment is quite a challenge. Not much is known about many species, and a lot of field work still needs to be done.

Animals have to be able to find each other to reproduce and establish territories. Since the individual animals may live quite far from one another, many have **calls** that can be heard for long distances. The rain forest is often a noisy place with all the calls of birds, insects, mammals, and frogs.

Some animals of the tropical rain forest live in just one layer of the forest. For example, the tapir, White-tailed Deer, and Brocket Deer are only found on the forest floor. The Black Howler Monkey, toucan, and the Harpy Eagle are almost always found in the upper canopy. A few animals travel through all the layers. The Coati, a racoon-like animal, can be found anywhere, and the Tree Boa, a large snake, can be found in both the middle and upper layers.

Sometimes animals are captured and fitted with a small radio transmitter. This allows the researchers to find the animals by radio signal. They can follow the signals to see how far the animals travel in a night. They can also learn how large an animal's home range is.

Animals may be wildly scattered, but they are very important to the plants of the rain forest. Very few rain forest plants are wind-pollinated, so animals such as monkeys, insects, bats, and birds carry on most of the pollination of the flowers. There are almost 1,000 species of fig trees, and they are pollinated by about 1,000 species of wasps. Animals also take on seed dispersal chores. A few very tall plants or the epiphytes of the upper canopy have small, wind-dispersed seeds, but most tropical rain forest plants have a seed with a fleshy coating. These seeds are attractive to birds, bats, and other animals. The fruit is eaten, and the seed is regurgitated later or passed through the digestive tract of these animals.

Name ______________________________ Date ______________________

For the Student:

1. Why is the Three-toed Sloth hard to see?

2. What are two animals of the tropical rain forest that are found in small herds or family groups?

3. Why are plants of the forest floor not wind-pollinated?

4. What is the largest cat species of the tropical rain forest?

5. Why is studying the animals of the tropical rain forest difficult?

6. What type of camouflage does a jaguar have?

7. List three animals found only on the forest floor.

8. List three animals that almost always stay in the upper canopy.

9. List two animals that can be found in several layers of the rain forest.

10. How do scientists keep track of some animals?

An Abundance of Birds

The tropical rain forest has a great number of different species of plants. It also has more species of birds than any other region of the world. Just how many species is not exactly known, because more of them are being discovered every year. In North America, north of Mexico, there are about 700 species of birds. In South America there are 2,900 species of birds. That's about one-third of all the species found throughout the world.

With all these species of birds, you might expect to be surrounded by birds on a walk through the rain forest; however, this is not the case. The rain forest holds a lot of different kinds of birds, but not huge numbers of any one species.

In the United States, you can see flocks of thousands of water fowl, like ducks or geese, or you can see hundreds of pigeons, starlings, or Red-winged Blackbirds. As a rule, huge flocks of birds are not seen in the rain forest. An exception to this rule occurs along a few mineral clay banks found along some of the rivers. Scientists believe that these minerals help neutralize the poisons found in some fruits that are eaten by parrots and macaws, so flocks of parrots and macaws, sometimes numbering in the hundreds, are found by these mineral banks.

Many of the birds of the rain forest are shy and secretive. You have a better chance of hearing their calls than seeing them. The most birds ever reported by bird watchers in one day was in a tropical rain forest. They identified more than 300 species of birds! This feat was accomplished on the Manu River in Peru. The bird watchers were experts on the calls of the tropical birds and most of the birds were identified by call and not by sight.

Scientists know a lot about the birds of the United States. Whole books have been written on birds like the Cardinal, Song Sparrow, and Bluebird. There is very little known about many of the birds of the tropical rain forest. Some of these birds have only been observed a few times. Some of the birds are fairly common, but their nests have never even been located or studied. Recently, a new species of bird never cataloged by science was found in Brazil. The surprising part of this discovery was that the bird was not hidden away in a remote section of the forest. It lived next to a busy, four-lane highway!

Name ______________________________ Date ______________

For the Student:

1. About how many species of birds are found in North America (excluding Mexico)?

2. About how many species of birds are found in South America?

3. What is the most species of birds reported in one day in Peru?

4. Where can you find large flocks of parrots and macaws?

5. Why are the birds at these locations?

6. Are you more likely to see or hear a bird in the rain forest? Why?

7. Where was a new species of bird found recently?

8. Are there usually large flocks of any one species found in the rain forest?

Name ______________________________ Date ____________________

Activity:

Listed below are some birds that are found in the tropical rain forest. Pick a bird from Column A and another one from Column B. Research both of the birds. Prepare a report describing what you learned about the birds.

A	B
King Vulture	Crested Guan
Sun Bittern	Nocturnal Curassow
Scarlet Macaw	Ruddy Quail-Dove
Long-Tailed Hermit	Scarlet-Shouldered Parrotlet
Green-Bellied Hummingbird	Gould's Jewel Front
Resplendant Quetzal	Toady Motmot
Teal-Billed Toucan	Scaled Antpitta
Great Antshrike	Purple-Throated Fruitcrow
Three-Wattled Bellbird	Red-Legged Honeycreeper
Harpy Eagle	Oilbird
Guianan Cock-of-the-Rock	

Foraging Flocks

Birds of the tropical rain forest have developed a number of interesting feeding strategies. Many birds travel in what we call a **foraging flock**. When we see a flock of birds in the United States, it is either one species or a few species mixed together. A tropical rain forest foraging flock of birds may be made up of 20, 30, or even 60 different species! There are usually only a couple of individuals from each species in the group, so the total number of birds is not large, but the number of species is.

An encounter with a foraging flock might go something like this. You have walked along a forest trail for 20 minutes without seeing a bird. Suddenly, you hear a loud chirp and a robin-sized bird flies across the trail. It lands on an exposed limb near the top of a canopy tree. Through the binoculars, you see a yellow bird with black wings, a black head, and a white throat. It's the White-throated Shrike-tanager. This bird is often a flock leader because it is noisy and alert. The soft chirps and twitter of more birds announce the arrival of the rest of the flock. They are spread out through the forest. Some birds are as high as the Shrike-tanager, and others are as low as the ground. They are covering all of the layers of the tropical vegetation.

A clump of dead leaves about 60 feet above the ground catches your eye. It's moving! Out pops a reddish, nondescript bird. It's a Ruddy Foliage-gleaner. As it probes the dead leaves for insects, a large katydid flies out. The Shrike-tanager has it in a second. After smashing it against a limb, he swallows down the large insect.

Now you hear something in the dry leaves of the forest floor—a Jaguar sneaking up on you? No, it is a Scaly-throated Leaf Tosser. It is looking for insects under dead leaves. Suddenly, birds are moving all around you. A black and white bird walks down a tree limb

looking for insects in the bark. This is an easy one. You see it in the United States every spring. It is a Black-and-white Warbler, a migrant bird spending the winter in the tropics. Another bird is moving up the trunk of a tree. Its big bill knocks off pieces of bark as it searches for insects. It is a Stony-billed Woodcreeper.

All of this action has a number of insects flying around, and there are birds in pursuit. A Sulfur-rumped Flycatcher and Tawny-crowned Greenlit seem to be catching their fair share of the bugs. Suddenly, you notice, just ten feet away on a branch near the trail, a robin-sized, black and white bird. The bird is not moving. It just sits there. A quick check in the bird book helps you identify it. It's the White Whiskered Puff Bird. This is an insect eater, but it prefers to wait for a big insect like a praying mantis. It does not move as you walk by.

A flash of bright blue in some yellow flowers catches your eye. Its red legs confirm that this is a Red-legged Honeycreeper hunting some nectar in the flowers. A beautiful, clear bird song starts on the left side of the trail. Suddenly, a bird joins it on the right. It's a pair of White-breasted Woodwrens singing a duet.

The birds are going by too fast, and you are missing some. There's a Plain Xenops acting like a Black-capped Chickadee. Several tiny Dotted-winged Antwrens flit around for a moment, and then they are gone.

It is getting quiet. You want to follow the flock, but they do not follow the forest trails. One last straggler comes into view. This bird is a very bright green with a bright yellow tip over the eye and a blue crown. You watch this last, beautiful bird fly off to join the flock. It is a Golden-browed Chlorophonia.

Ornithologists, scientists who study birds, are beginning to understand why these diverse birds travel together in a flock. The flock leader, the Shrike-tanager, is very alert and loud. A Foliage-gleaner is easy prey to a hawk when its head is buried in leaves, but the Shrike-tanager's loud call will warn it if a hawk is spotted. The birds have different feeding techniques and eat different-sized insects and fruit. Insects one bird stirs up may be food for another bird. The foraging behavior is beneficial for all the birds in the flock.

Name ______________________________ Date ____________________

For the Student:

1. How many birds from each species are in a foraging flock?

2. How many different species may be in a foraging flock?

3. Why would you call the Shrike-tanager the flock leader?

4. How does a Leaf-tosser find its food?

5. What bird knocks pieces of bark off the tree as it searches for insects?

6. What bird searches for nectar in flowers?

7. What is an ornithologist?

8. Why do birds of different species travel together in a foraging flock?

Ants and Plants

I was bird watching with friends in Belize. We were staying in a lodge in the tropical rain forest. Next to the walk of the lodge, I noticed a beautiful Croton bush. Its leaves were a bright yellow with red stripes. The next day when I walked by, I noticed that the leaves were gone, every one of them! Leading away from the bush was a new, foot-wide trail. The trail was bare of all living vegetation, but here and there I saw little finger-sized pieces of Croton leaves. I followed the trail into the forest and came upon a group of mounds—some small, some quite large—with entrance holes in them. It was the hottest time of the day, and nothing was moving, so I returned to the lodge.

The next day, I went back early, and the ground was alive with ants. The ants were traveling on trails that radiated out from the mounds. The ants that were going out were not carrying anything, but the returning ants each held up a small piece of leaf. Marching with these worker ants were soldier ants. The soldier ants were larger and had big mandibles, or pinchers. There were guarding the workers.

The Leaf-cutter Ants weren't taking the leaves into their mounds to eat. Instead, they were using the leaves as nourishment for a special fungus. The ants find food for the fungus and then eat the fungus themsleves. Without the fungus, the ants would die, and without the ants to provide the leaves, the fungus would die.

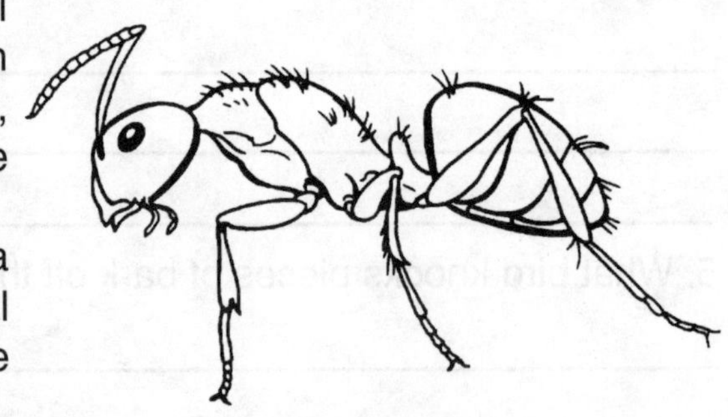

Some of the plants in the rain forest have a chemical in their leaves that acts as a natural fungicide. The chemicals kill any and all fungi. The ants don't harvest any leaves from those plants.

Another type of ant lives in the hollow thorns of a small tree. These ants are fierce little guys, and they will attack any animal that tries to eat the plant in which they are living. The tree provides a home for the ants, and the ants provide protection for the tree.

Another group of ants doesn't have a home. They travel along, carrying the queen, the eggs, and the young ants with them. These ants are called Army Ants. They travel in a swarm that may be 20 feet wide. They capture and devour all the insects, frogs, lizards, and small snakes in their path. As the swarm of Army Ants travel, insects and other small animals flee in front of the swarm. A group of birds called Antbirds forage in front of the ant swarm, catching the fleeing insects. There are over 100 species of Antbirds in the tropical rain forest. Army Ants help them find their food.

Ants are just one of the thousands of groups of insects found in the tropical rain forests. They have an important role in the intricate web of life in the rain forest.

Name ______________________________ Date ____________________

For the Student:

Read the following sentences. If a sentence is true, circle the "T." If a sentence is false, circle the "F."

T F 1. Ants are always harmful to plants.

T F 2. Leaf-cutter Ants only eat certain leaves.

T F 3. Leaf-cutter Ants farm fungi.

T F 4. It is only in the movies that Army Ants attack and devour people.

T F 5. Most species of ants have soldier ants that can bite.

T F 6. Leaf-cutter Ants can tell the difference between leaves that will grow fungi and leaves that won't.

T F 7. The queen of the Army Ants lives in a large ant hill.

T F 8. Ants that live in hollow thorns eventually eat the trees they live in from the inside out.

T F 9. Army Ants help Antbirds find food.

T F 10. Antbirds follow the Army Ant swarm to find insects left by the ants.

Find and circle the following words in the word search puzzle below. Words may be printed forward, backward, horizontally, vertically, or diagonally.

1. ANTBIRDS
2. ARMY ANTS
3. BELIZE
4. FUNGI
5. LEAF CUTTER
6. QUEEN
7. SOLDIER
8. SWARM
9. THORNS
10. WORKER

U W I Q T U B Q L L H F X W V I G F B W
K U R H G H H O K Q V G H X G M G M Z S
A K J A I K T U B J D Y W N P M P B N Y
S D E K H W G P W S A C U H B R S R I S
X C A K Q J F G M J J F S R M S O A P I
C N N G D G A B W U M H W E F H L X J W
M Q T N L P B R H Z W R K I T A D S F H
L U B G J M Q U M O V V A D C E I L M Z
R Y I L U M G W Z Y J B A W Y V E T O T
Q L R B M T N O U H A F A E S A R Q N Q
H Q D I D W Q R H L D N V A I T C S D M
H Y S J D U O K C H V J T M B P F I C P
A M M A E B H E M X U B C S Q B O O F V
Y Q V E Q L N R A L E A F C U T T E R G
E P N G U U F G B U C Z Q S Y A A Z P G
E J B B N Q V Q Q G J P W E K J N G A E
B A E W Q Z O R L J Z X K B K Z K Z S Y
D U L N X M B E L I Z E U N T J O S G X
M C F F U W T W T C U U X O Q T F R D W
S M Z O V G F T V S X Q H B H Z N D S A

Hot and Humid, Dark and Dangerous

The tropical rain forest has been protrayed in books and films as a very unpleasant place. In the movies, the hero is often overcome by the heat, mired down by quick sand, covered by huge biting insects, and attacked by a variety of animals ranging from electric eels to jaguars to cannibals.

Tropical rain forests are warm, but the temperatures rarely reach 90 degrees Fahrenheit. There is also a lot of rainfall, but during the dry season, days go by without any rain.

Many people have found bothersome insects, such as mosquitoes and biting flies, to be much more numerous in our own temperate zone than in the tropics. Some tropical insects are huge and fierce looking, but most of them are not harmful to people.

Snakes are not usually encountered. Some poisonous snakes are found in the rain forest, but you would probably have to look very hard to find one.

Jaguars, ocelots, and pumas are predators, but attacks on people are very rare. You can consider yourself lucky if you ever even spot one of these cats.

Piranha have very sharp teeth, but they rarely attack unless there is already blood in the water. Electric eels, freshwater sting rays, and caiman (tropical alligators) are easily avoided.

There are some poisonous plants in the tropics. One of them is called the Poison Wood Tree. Local legends say that if you even walk in the shade of it, you will break out in a rash. The real story is that some people will break out in a Poison Ivy-like rash if they come into contact with the leaf or sap of this tree. Poison Ivy is much more common in temperate forests than the Poison Wood Tree is in tropical rain forests.

A very strong poison is found in the skin of some brightly colored tree frogs. Some natives still use this poison on their arrows or darts. There are very few hostile tribes; however, and they are found in the most remote areas of the forest.

So, travel in the tropical rain forest is easy, comfortable, and completely safe. Well, not exactly ... yellow fever, tyfus, malaria, hepititis, and other diseases are more prevalent in the tropics. You can get shots and take precautions to minimize your risk. Internal parasites transmitted by insects can be a problem, especially for those who spend long periods of time in the rain forest. There are, however, very effective treatments for most of these parasites.

Don't pass up the chance to visit the tropical rain forest because of creative stories and films. It is a very interesting area of the world in need of much more scientific study.

Name ____________________ Date ____________

For the Student:

1. Which place gets hotter in August: St. Louis or the Amazonian forest?

2. Are the most dangerous animals of the rain forest large or small? Why?

3. Where do the natives collect the poison for their arrows?

4. What is a caiman?

5. Can you get a rash from the Poison Wood Tree without actually touching part of it?

6. What causes the rash from the Poison Wood Tree?

7. What are three types of large cats found in the tropical rain forest?

8. What are four diseases that are more prevalent in the tropics?

9. How can you avoid getting these diseases?

10. Do people usually encounter poisonous snakes in the tropical rain forest?

Name ______________________________ Date ____________________

Activity: Categorizing Organisms

Research one of the following animals. Explain how your animal fits into life in the rain forest. (Example: It is a predator or prey species.)

Tapir

Capybara

Aguti

Oscelot

Jaguar

Piranha

Tyra

Anteater

Harpy Eagle

Sloth

Howler Monkey

Flying Fox

Chimpanzee

Duiker

Water Buffalo

Okapi

Iguana

Jungle Fowl

Chameleon

Boa Constrictor

Giant Otter

African Gorilla

Poison Arrow Frog

Bongo

Vampire Bat

Toucan

Marmoset

Pygmy Hippo

Brocket Deer

People of the Rain Forest

For thousands of years, people have lived in the rain forests. No one seems sure how these people got there in the first place, but we are beginning to understand how they survive. They usually live in small groups as **hunter/gatherers** or **farmers**. The hunter/gatherers hunt the animals of the forest for meat and gather forest fruits and plants for the rest of their food. Traditional hunting methods are used and are handed down through the generations. They use bows and arrows, spears, blow guns, and traps to kill their prey. They use the resources of the forest to build their huts. They know which forest plants can be used to treat their illnesses. When animals in one area become too wary or scarce, the hunter/gatherers move along to another area in the forest.

The rain forest farmers depend on the plants they have put into the ground for the bulk of their food. They hunt and collect some wild plants and animals, but they are more set in their ways. They practice **slash-and-burn agricultural techniques**. First, trees are cut down and left on the forest floor to dry. When the plants are dry enough, a fire is started. The ashes that are left fertilize the thin soil. A crop can be planted in the cleared and burned area. After a couple of years, however, the soil is depleted of its nutrients, and the field is abandoned. Soon the area will grow up again in trees. After about 20 years, the same area may be cleared and turned into a field once again. This practice is not detrimental to the rain forest because it is carried out on such a small scale and in such a small part of the forest.

Populations of people living in the rain forests have been declining. The isolated tribes have developed a resistance to parasites and tropical diseases, but they have no resistance to European or North American diseases, such as measles and tuberculosis. Even a common cold can cause death for the forest people. There may have been one million Indians living in the Amazon forest in 1900. It is estimated that the population is now down to about 200,000 individuals. Eighty-seven entire tribes have died out since 1900. The surviving rain forest tribes speak as many as 170 different languages.

Along with the **Amazonian Indians**, there are also the **Mayan Indians** living in the rain forests of Central America. The **Pygmies** live in central Africa, and the forest people of Borneo are called the **Penau**. In Thailand, the forest people are the **Lua**.

As the rain forests are depleted and changed, the people of the forest are under pressure to change, too. Some of them gather rain forest resources, such as Brazil nuts, coffee, and bananas, to trade for outside goods and foods. Others retreat farther into the remaining forest, trying to maintain their traditional life styles.

Name ______________________________ Date ______________

For the Student:

1. What are the names of three different groups of people who live in the rain forests?

2. What did the traditional hunter/gatherers use as weapons?

3. What is slash-and-burn agriculture?

4. How long does it take for a cleared plot of ground to become covered with trees and ready to be farmed again?

5. Why have native populations declined?

6. How many tribes have died out since 1900?

7. Why do rain forest farmers travel less than hunter/gatherers?

8. What products do some rain forest people trade for outside goods and foods?

9. What products do hunter/gatherers get from the rain forest?

Lost Cities

The tropical rain forests of the earth are the least studied and least explored regions of the terrestrial world. Wonderous stories and legends have grown out of these unknown regions.

In 1541 Francisco De Orellana, while sailing up a great river, reported the existence of fierce female warriors. The river became known as the Amazon, named for the female warriors of Greek mythology. More than 450 years later, we are still searching for lost cities, lost tribes, and new animals.

For many years Colonel P. H. Fawcett explored the interior of Bolivia and Brazil's rain forests. Indians of remote tribes told him of lost cities hidden in the forests. His expeditions penetrated far into the interior, but we do not know if he ever found any of the cities. He and all of his men disappeared without any trace in 1924.

An early 1990s expedition, sponsored by the National Geographic Society, went into the rain forest of the Congo River basin in Africa. They were searching for forest elephants and the **modele-mbembe**, a rhinoceros-like animal that was said to be the last living dinosaur. The dinosaur was never found, but the Pygmies that the explorers met said the dinosaurs did indeed exist. They called it the elephant killer.

Recently a goat-like animal was discovered in an Asian rain forest, and several new species of monkey were found in South America. Reports of a new species of a jaguar-like cat have not yet been confirmed. Natives claim a cat with exposed teeth, much like the extinct Sabre Tooth Tiger, still roams the rain forest.

On the Amazon River, one explorer claims to have shot a 65-foot long snake. The largest snakes ever actually captured and photographed have been more like 20 to 25 feet long.

As we learn more and more about the rain forest, many legends have been proven to be just stories. However, the stories of lost cities have proven to be true. When the first Europeans set foot in the "new" world, they found tall buildings towering over the dense vegetation. Most of these cities were deserted, and the rain forest was overgrowing them. Who built these cities and how?

We now know something about the "lost" civilizations that built these cities. The Mayas, for example, were flourishing in 500 B.C. They developed a numerical system based on zero. At that time, the Roman numerical system was in use in Europe. The Mayas had also developed a calendar with 365 days, just like the one we use today. Huge ceremonial cities were built. They were cities of cut stone with majestic palaces and tall temples. The tallest building in the country of Belize is still a Maya temple.

Scientists have found great stone monuments called **stelae** with carvings. The carvings were symbols telling stories of the king's personal accomplishments as well as the city-state's accomplishments. How were those stones, some weighing more than 20 tons, moved many miles through the forest? The Mayas didn't use wheels, and they did not have any draft animals. How did they cut the stones without any metal tools?

An even greater mystery is how these people survived and flourished surrounded by the thin soils of the rain forest. How did they grow their corn? Finally, what caused the sudden collapse of their empire?

Today the country of Belize supports about 200,000 people. Many of them are descendants of the Mayas. Scientists estimate that 1,000 years ago, the same country supported 2,000,000 Mayas.

New discoveries are being made every day, but still more research is needed. Maybe the "ancients" knew how to live in the rain forest without destroying it. Perhaps it is a lesson we need to learn again.

Activity:

Find out more about the lost cities. Each student or team of students should choose one city from the list below. Research the city and then prepare a written or oral report for the rest of the class.

1. Tulum
2. Uxmal
3. Palenque
4. Bonampak
5. Tikal
6. Copán
7. Chichen Itzá
8. Coba
9. Mayapan
10. Caracol
11. Altun Ha
12. Angkor

Name ______________________________ Date ____________________

For the Student:

1. Where did the Amazon River get its name?

2. What happened to Colonel P. H. Fawcett's expedition?

3. What do the Pygmies of the Congo River basin call the modele-mbembe?

4. How long are the largest snakes actually captured and photographed on the Amazon River?

5. Who built the great ceremonial cities in Belize and other Central American countries?

6. What was inscribed on the stelae of the Maya?

7. What do we know the Mayas did not use to move these huge stones?

8. How many people does the country of Belize support today? How many did it support in the time of the Mayas?

Activity: Farming, Jungle-Style

Dear Student,

We have just received a letter, and it is addressed to you. It came to us from a lawyer in La Paz, Bolivia, representing the estate of your late Great-uncle Herbie. Please read the letter and list of suggestions. When you have finished, please let us know what you would like to do. We will forward your final decision to the lawyer in Bolivia. We may be reached at:

Pat and Barb Ward
Tropical Rain Forest Land Use Consultants
R. R. 2, Box 138A
Murrayville, IL 62668

Hello kid!

You don't know me, and I never met you, but you and I are related. In fact, you are my only relative, so if this malaria gets any worse, or I just get old and fall over dead, I want you to have my ranch and take care of the three Indian families living on it.

Now, the ranch is a thousand acres of rain forest. We have been able to pay the taxes by collecting Brazil nuts and selling natural rubber collected on the surrounding public lands. We do a little slash-and-burn agriculture and some hunting for food. However, a new road has been built, and all of this has to change. People have settled on the surrounding land and are cutting down all of the trees. We will no longer be able to collect Brazil nuts, and the hunting is no longer good.

I've thought and thought about what to do, and I have even made a list of possibilities, but I'm old, and I expect you'll have to make the final decision. I hope whatever you decide, you will still be able to support the 20+ Indians and protect as much of the forest as possible. Do some research and give it some serious thought. It's up to you. The ranch is 90 miles southwest of Santa Crews. You'll recognize it by the huge Mahogany tree next to the pretty river. They'll bury me underneath that tree.

Trusting in you,

Great-uncle Herbie

Name ______________________ Date ______________

Activity: Farming, Jungle Style

Herbie's Suggestions

1. Extend the slash-and-burn agriculture to 100 acres a year.
2. Plant coffee and cashew trees.
3. Flood the area and grow rice.
4. Cut down all of the Mahogany and Rosewood trees for timber.
5. Teach the Indians to make bamboo furniture and other crafts.
6. Turn the ranch house into tourist accommodations for birdwatchers and photographers.
7. Clear the land and start a banana plantation.
8. Start a cattle ranch.
9. Develop a citrus plantation.
10. Start a houseplant and orchid nursery with plants collected from the rain forest.
11. Sell the land and tell the Indians to hit the road.

After researching the possibilities above or coming up with your own idea, write a letter to the land use consultants outlining your plans.

Great Rain Forest Drains and the Amazing Amazon

Tropical rain forests receive at least 60 inches of rainfall per year. Plants in the forests soak up some of this moisture, but some **runoff** also occurs. The runoff starts as rivulets, then builds to become streams, then ends as rivers—in fact, as mighty rivers.

The great rain forests of South America are drained by the Amazon River. The Congo River drains the forests of west central Africa, and the Nile begins its journey in the cloud forests of Africa. Each of these rivers is longer than the Mississippi.

The Amazon River discharges more than 7,000,000 cubic feet of water per second. The Mississippi discharges 651,000 cubic feet per second. The discharge of most rain forest rivers varies with the upstream rainfall. This is especially evident with the Amazon River. If you could measure all of the water in all of the rivers around the world, at times one-fifth of that water would be in just one river—the Amazon. All the rest of that water would have to be divided up among all the rest of the rivers of the world. The Amazon is so large than an island at its mouth, called Marajo, is larger than the country of Switzerland!

During the wet season, the Amazon River covers a huge flood plain area. Many plants and animals have adapted to these dramatic changes in water levels. Trees put on huge amounts of seeds. These seeds are dispersed by the flood waters so new trees have a chance to grow in areas away from the dense shade of the rain forest. Fish feed and spawn in the newly flooded areas, and small fish find protection from predators in the flooded vegetation. A layer of fertile silt is deposited by the floods, and this adds nutrients to the shallow rain forest soil.

The Congo River is unusual because it wanders both north and south of the equator. When it is the dry season north of the equator, the wet season is in full swing south of the equator, and vice-versa. In this way, the Congo River is always getting rain somewhere. It can maintain an almost constant flow. The flow is so strong at the river's mouth that, much like the Amazon, no real delta forms. Silt is carried far out into the ocean where it settles to the sea floor.

Name ______________________________ Date ________________

For the Student:

1. What are three great rivers that drain rain forests?

2. Why doesn't the Congo River have seasonal flooding?

3. Does the Mississippi River or the Amazon River discharge more water?

4. Why doesn't the Congo or the Amazon River have a large delta like the Mississippi?

5. Are Amazon floods destructive for all plants and animals? Why or why not?

6. How much of the world's river water flows through the Amazon River at its peak?

7. At least how much rainfall occurs each year in a rain forest?

8. What is Marajo?

Activity: Endangered Species, Problems and Solutions

Background Information

One hundred years ago, ladies wore hats when they went out in public. It was considered very stylish to have feathers on the hat, and not just any feathers would do. Colorful feathers, such as those of the Cardinal, Goldfinch, parrot, and even the hummingbird, were in great demand. Gracefully flowing plumes, such as those found on herons, egrets, and spoonbills, demanded a high price. Sometimes, hats were adorned with whole mounted birds.

People could make good money hunting birds for feathers. It was said that egret plumes were worth their weight in gold. With all of the hunting pressure, bird populations dropped. Egrets and herons were especially hurt because they were hunted at their nesting colonies. When the adult birds were shot, the young birds starved to death. Bird protection laws were finally passed, attempting to put an end to the problem.

The Audubon Society hired wardens to protect the remaining nesting bird colonies. In the Florida Everglades, one warden was shot to death by poachers. Finally, with new laws and a change in hat styles, the bird populations were able to recover.

Perhaps you are thinking, this kind of thing could never happen today, but once again, styles are causing the decline of some bird species. In fact, some birds have been listed as endangered. What is the problem this time? The problem is that we like to have pretty, talkative birds in our homes in cages. Parakeets, parrots, cockatoos, macaws, and a long list of other birds are considered to be fun pets.

We have laws that say that it is illegal for us to keep a native bird as a pet, so you don't see Cardinals or Goldfinches in cages. The law doesn't extend to birds from other countries, however. We can have birds for pets, as long as they come from another country. The rarer the bird is, the more it costs, so some of the native people of the rain forest are catching rare birds to sell. It's hard to catch the adults, and the young make the best pets, anyway. Nest trees are cut down, and the young who survive the fall are captured. They are kept in cages until they can be shipped out of the country. Most of the time, they are not well cared for, and many of them die.

New laws have been passed making it illegal to import endangered species. Birds such as the beautiful Hyacinth Macaw are very rare in the wild. A macaw might sell for several thousand dollars. That's a lot of money, especially for a poor local farmhand. So, despite the laws, the birds are still being captured. There is some work being done to breed these endangered birds in captivity so the young can be released back into the wild. Each endangered species has its own set of problems, and plans have been developed to protect them.

Pick out an endangered species from the list below. Find out why it is endangered and what is being done to protect it.

Asian Tiger	Lemur	Golden Cat	Indian Python
Scarlet Macaw	Giant Otter	Hyacinth Macaw	Black Rhino
Orangutan	Pygmy Loris	Ocelot	Golden Line Tamarin
Aye-aye	Harpy Eagle	Okapi	Yellow-Headed Parrot
Lowland Gorilla			

Cycles and the Rain Forest

Water is found in many different forms. It is found as a liquid in our rivers, lakes, ground water, and oceans. It is also found as a solid in ice and snow. A huge amount of water is tied up in the earth's polar ice caps. Water can also be a gas or vapor found in the atmosphere, sometimes in clouds.

Water changes among these forms, but the actual amount of water on Earth is a constant. The water you drink today may have, at one time, been released into the air by a rain forest tree as water vapor. This atmospheric moisture could later fall as rain, replenishing your town's water supply. Let's see in a bit more detail how this **water cycle** works. Trees in the tropics shade the ground and protect it from the Sun and dry winds. The roots of these trees take up rain water. The water travels through the trees' tissues and evaporates from the trees' leaf surfaces. This moisture forms clouds, and eventually falls to the earth as rain.

When these trees are cut down, the water cycle can change. Dry seasons become even drier and hotter. Since there are fewer roots to take up the moisture, runoff into streams and rivers is accelerated. Less water is found in the topsoil and the ground water level drops. Droughts and floods can become common occurrences.

Another cycle that is affected by the rain forest is the **carbon cycle**. Huge amounts of carbon are tied up in the wood and leaves of the rain forest. As these plants grow, they absorb carbon from the air in the form of carbon dioxide. They use this carbon dioxide to make their own food in the process known as photosynthesis. When making their own food, plants also make extra oxygen, which they release into the air. When large numbers of trees are cut down and burned, less carbon dioxide is absorbed from the atmosphere and less oxygen is produced. The amount of carbon dioxide in the atmosphere has been increasing by about four percent a year.

The increase in carbon dioxide is caused by the burning of fossil fuels, such as oil, coal, and natural gas. It is also caused by the slash-and-burn clearing of large areas of rain forest. Extra carbon dioxide in the atmosphere causes a condition that scientists call the **greenhouse effect**. Solar radiation from the Sun's rays warms the earth's surface. This warmth is radiated back into the atmosphere. Have you ever noticed how much cooler it gets on a clear night than on a cloudy night? When there are clouds, the radiated heat is reflected back towards the earth. It is trapped by the clouds and is unable to escape into the atmosphere.

Gases such as carbon dioxide, methane, chloroflourocarbons, and nitrous oxide can all contribute to the greenhouse effect. They help hold the heat close to Earth. It is estimated that the worldwide temperature rise will be between four and 11 degrees over the next 100 years. If this happens, the climate will change throughout the world. Crop production will be altered, and the ice caps will begin to melt. The ice cap melt will cause the seas to flood large areas that are currently dry land. This will affect millions of people in a very negative way. Protecting our rain forests is one way to help protect our environment.

Name ______________________________ Date ____________________

For the Student:

1. What are states of matter in which water is found naturally in the world? Give an example of each.

2. How can increased heat in the atmosphere cause flooding?

3. How does excessive carbon get released into the atmosphere?

4. Scientists think that global warming may increase the world temperature. What is the expected increase?

5. How could global warming affect non-coastal areas?

6. Complete the graphic organizer below illustrating the carbon cycle. Use the terms below to fill in the blank boxes.

Trees

Take In Carbon Dioxide

Release Carbon Dioxide When Burned

Release Oxygen

Name ______________________ Date ______________

Activity: Products of the Rain Forest

All of the following products originally came from areas in or near the rain forest. Take a look at the list. Make a check mark next to all of the products that you, or your family, use.

___ 1. bamboo	___ 2. sesame seeds
___ 3. balsa wood	___ 4. sugar
___ 5. teak furniture	___ 6. tapioca
___ 7. sandal wood	___ 8. tea
___ 9. rose wood	___ 10. vanilla
___ 11. mahogany	___ 12. turmeric
___ 13. jute twine	___ 14. paprika
___ 15. burlap	___ 16. nutmeg
___ 17. rattan furniture	___ 18. mace
___ 19. kapok life jackets	___ 20. ginger
___ 21. chickle (chewing gum base)	___ 22. cloves
___ 23. copal varnish	___ 24. cinnamon
___ 25. rubber balloons or balls	___ 26. chocolate
___ 27. palm oil	___ 28. chili pepper
___ 29. coconut	___ 30. cayenne pepper
___ 31. camphor	___ 32. cardammon
___ 33. limes	___ 34. black pepper
___ 35. Brazil nuts	___ 36. allspice
___ 37. cashews	___ 38. avocados
___ 39. coffee	___ 40. bananas
___ 41. cola	___ 42. grapefruit
___ 43. corn	___ 44. guavas
___ 45. macadamia nuts	___ 46. palm heart
___ 47. peanuts	___ 48. lemons
___ 49. rice	___ 50. mangos
___ 51. oranges	___ 52. papayas
___ 53. passion fruit	___ 54. pineapples
___ 55. plantains	___ 56. potatoes
___ 57. sweet potatoes	___ 58. tangerines
___ 59. tomatoes	___ 60. yams

Our Most Important Forest

For people living and working in a big city in California, farming on the great plains of Kansas, or vacationing in the deserts of Arizona, the tropical rain forest seems to be a long way away. They might not consider these faraway forests to be important, but they certainly are. Let's look at the facts.

Tropical rain forests are an ancient type of plant community. They have been developing for more than 30 million years! This slow development has created an incredible variety of forms of life. Scientists believe that more than half of all plant and animal species on Earth are found in the tropics. Many of these life forms are so specialized that they can only live in an intact rain forest.

Plants and animals of the rain forest provide the basis for many foods, drugs, and other products that we use every day. New plants and animals are being discovered, but some vital plants and animals are lost forever by the destruction of the rain forests. The loss of these undiscovered plants and animals means the loss of potentially useful products.

The rain forest also provides stability in the climate. The plants of the forest take up rain water through their roots, and it later evaporates from their leaves. This process cools the air and adds moisture to it. When rain forests are cut down, the local climate can change. Some areas have experienced hotter temperatures and periods of drought. The loss of huge areas of the rain forest could effect the climate all around the globe. In other words, the rain forest can affect your weather, no matter where you live.

The rain forests also carry on a protective function. They protect the thin soil from washing away in heavy rains. They slow the runoff of rain water so that river systems are not clogged with silt. This can have a far-reaching benefit. A large number of sea animals begin life in **estuaries**, areas where fresh water meets salt water. Esutaries are often found where rivers meet the sea. If these waters become polluted with silt from rain runoff, they are not as good for marine life. The sea animals depending on the estuaries will suffer. Slowing the runoff of rainfall also reduces flooding. It also adds to the amount of available ground water. In Asia, villages have had all of their wells run dry after forests were cut down.

Many birds that we see in the United States spend their winters in the tropical rain forests. As rain forests are cut down, these bird populations have declined. Insect-eating birds are important to our temperate forests and even help the farmers by eating large numbers of insect pests. If the birds have no place to go in the winter, they will not return to our area in the spring. They will not eat up insects, and we will suffer.

The tropical rain forests are still home to a number of native people. These people have a lot of knowledge about local plants and animals and their uses. The homes and ways of life of these native people will be destroyed if the rain forest disappears.

Activity:

After the class is divided into groups of four or five students, each group should pick out at least two reasons why the rain forest is important. You may think of even more reasons. List all of the reasons on the board. The class can rank the reasons according to how important they are to the students themselves. Rank the most important reason as number one, etc.

Name ______________________________ Date ______________

For the Student:

1. Where are more than half of the plant and animal species on Earth located?

2. How long has the rain forest been developing this great diversity of life?

3. What is an estuary?

4. How does the rain forest protect an estuary?

5. How do rain forests cool the air and add moisture to it?

6. How does cutting the rain forest affect birds in the United States?

7. How do decreasing bird populations hurt farmers in the United States?

8. What happens to native people when rain forests are destroyed?

The Bush Doctor and Shopping in the Rain Forest Drugstore

Natives living in the rain forest are often a long way from modern medical treatments. Over hundreds of years of trial and error, they have figured out which native plants have medical uses. The rain forest has become their drugstore. Each tribe has a **medicine man**, sometimes called a **witch doctor** or **bush doctor**. This person has been taught about the medicinal plants, and it has become his duty to teach the next young apprentice. In this way, the knowledge is passed from generation to generation.

One warm day, a group of us were hiking a rain forest trail in Belize. We came to a small clearing with a thatched-roof hut. This was the home of a bush doctor. She was a typical-looking Mayan Indian, dressed in a blouse and skirt—not a feather headdress and magic wand.

The bush doctor looked over the group as we walked up. She quickly addressed one man in the group, saying, "Oh, so you are a diabetic."

I don't know how she knew. He looked fine to me. She took out a jar of bark. "Drink a tea made from this every day, and you will not need insulin after only three months."

I asked the bush doctor if she knew a cure for asthma, a problem my daughter has been suffering with for many years. "Well," she said, "That's a tough one. I cured a niece, once. You have to catch a jungle rat. You cook it whole, and she will have to eat it all, even the bones."

I didn't know what would be more difficult, getting a jungle rat home, or getting my daughter to eat it. She invited us to join her, so we drank a cup of Blood Tea, made from the sap of a jungle vine. It was very refreshing, and she said it was also good for the circulation. As we looked around, we noticed the bush doctor had a number of different herbs drying and several small piles of bark. In jars she had lots of different roots. My friend bought the bark for his diabetes, and we said goodby to the bush doctor.

I can't say that the bark worked for my friend; his wife didn't trust it, and she threw it out. I never did get a jungle rat home, either. However, the rain forests have produced over 5,000 products that we use today, some examples are listed below.

Quinine: From the bark of the Cinchoma tree, this is used to treat malaria, and is used in some carbonated beverages.

Reserpine: From Rauvolfin plants, this is used to treat high blood pressure.

Vincristine: From the Rosy Periwinkle, this is used to treat Hodgkin's Disease and Leukemia.

Ipecac: From roots, this is used to treat dysentery.

Curare: From the sap of a vine, this is a muscle relaxant.

Rotenone: From roots, this is an insecticide.

Dragonsblood: From the sap of the Swampkawg tree, this is used to treat various blood disorders.

Balache: From the Dogwood tree, this is an intoxicating beverage.

Billywebb Bark: From the Billywebb tree, this is used to treat coughs and fever.

Cosmetics: Special oils from the Copaiba tree are used in cosmetics.

Incense: This comes from the sap of the Qunbo Limbo tree.

Annatto: Seeds from this tree produce an orange dye used in butter, cheese, varnishes, etc.

Many plants have more than one use. Probably, the "king of useful rain forest plants" is the Cohune Palm. It's a very common tree, and is called the tree with 1,001 uses. These uses range from edible nuts and palm oil to roof-thatching material and charcoal.

Name ______________________________ Date ____________________

Bush Doctor Word Search

Find the words listed below in the word search puzzle. Words may be printed forward, backward, horizontally, vertically, or diagonally.

Y R E S E R P I N E W O U C P M U M Q C

Y N F G E H C A L A B Z C R J N K P M X

C M N M F I Z X Y H F I N C E N S E O I

I Q P W M E D I C I N E M A N C R S T B

V R N O L B B B S K H T O S L L V O T S

M I O E L K N I W I R E P Y S O R J A K

I V H T Q U N B O L I M B O P M A V N K

P I I H C H M Z O M E T F V I L Z Q N A

J A G N J O I E H G D N E X T L D Y A H

Z Z C C C H D W I N W O O I K D T U I T

J C J O R R X H X L X A G N P N R E L A

W X J H P A I P S J M E K W E E U Z K R

P M X E O A V S Y U O M Y P O T C T R E

S N X L T O I O T R B P L Z M O O A I L

Z W Z P M F Y B B I K X P K O A D R C G

E K O Z N I U G A B N C X L V X W Z G N

I C I N C H O M A M Y E P O A H H S L U

O E B I O D O O L B S N O G A R D X K J

L G W T K B I L L Y W E B B B A R K U N

D Y B L Q U I N I N E J V G E R A R U C

ANNATTO
BUSH DOCTOR
CURARE
INCENSE
MEDICINE MAN
RESERPINE
SWAMPKAWG
BALACHE
CINCHOMA
DOGWOOD
IPECAC
QUININE
POSY PERIWINKLE
VINCRISTINE
BILLYWEBB BARK
COPAIBA
DRAGONSBLOOD
JUNGLE RAT
QUNBO LIMBO
ROTENONE

Ecotourism

Everyone knows what a **tourist** is. It is someone who travels for pleasure. Maybe you have been a tourist. Have you traveled to an amusement park or gone on a fishing trip or visited some historical site? Most of us, at one time or another, have been a tourist.

There is a special type of tourism that is becoming more and more popular. It is called **ecotourism**. Ecotourism is travel for pleasure, but it is more than that. It is travel to see some aspect of the natural modern world. Birdwatchers traveling to see some rare bird are ecotourists. Maybe you have been an ecotourist. Have you ever traveled to see mountains, the Grand Canyon, or an ocean?

So why is ecotourism important to the rain forest? The rain forest has a great diversity of life forms. There are lots of birds, butterflies, flowers, trees, and animals. There are a lot of people interested in seeing those birds, while others prefer the butterflies or orchids and tropical trees. Some people want to see the whole system. This interest by ecotourists is being used to protect some areas of rain forest.

In the country of Belize, the owner of a large tract of rain forest was having big trouble with trespassers. They were shooting wildlife and digging up some ancient Mayan ruins looking for artifacts to sell. The owner thought about hiring guards, but it was a large area, and it would have been very expensive to pay the guards. Instead, the owner came up with the idea of building an ecotourist lodge near the Mayan ruins.

His plan worked very well. The lodge is called Chan Chich, which means "little bird" in Mayan. The lodge has several miles of hiking trails through the rain forest and around the ruins. Local people from a nearby village work at the lodge and serve as tour guides. The tourists have spent a lot of money in the area, and the village now has a school because of the increase in their economy. The local villagers have noticed that people will travel very long distances and spend a lot of money to see the intact rain forest. Now the local people are beginning to understand that protecting the forest will support them far long than cutting down the trees. The cut trees will only give them a short-term income.

Another example of the benefits of ecotourism can be found in Panama. There the Kuni Indians have been trying to protect the land that the government gave them in 1925. Originally, the land grant was hundreds of thousands of acres. Today, their rain forest area has been reduced to about 60,000 acres. The rest was taken from the Indians by force, and the trees were cut down for timber. To protect the last of their rain forest, the Kuni have built a lodge for ecotourists. They are using the money that they make at the lodge to put up signs. Since the government wants to encourage tourists to come to their country and spend lots of money in the country, the government is finally helping the Indians protect their land.

In Costa Rica and Belize rain forest tourism has made a big difference in the economy. Both countries are setting aside large forest preserves hoping to attract as many tourists as possible.

Other countries are following their example, and ecolodges can be found from the depths of the Amazon to the mountain slopes of New Guinea. International conservation groups, such as The World Wildlife Fund and the Nature Conservancy, have provided financial assistance for these new protected areas.

Name ______________________ Date ______________

For the Student:

1. What is a tourist?

2. What is ecotourism?

3. Have you ever been an ecotourist? If so, list some of the places you have visited.

4. How do natives of the rain forest try to protect their forests using ecotourism?

5. List three countries where rain forests are being made accessible to ecotourists.

Class Activity:

Draw up a plan to build an ecotourism lodge in the center of a 250,000-acre rain forest park. Keep in mind the comforts that most ecotourists enjoy. For example: a roof over their heads, three meals a day, clear paths, electricity, flush toilets, and so on. Develop a list of materials that will be available in the rainforest. Make a separate list of the things that you will have to bring in from other areas.

Down the Napo River

The opportunity to visit a remote area of a tropical rain forest was too good to pass up. John, the owner of a birdwatching tour company, had called with information. The Indians of the area had built a lodge and were looking for some income from tourists. They were also hoping that the area would be made into a park so it could be protected from loggers. John asked if I would take a group in and check out the accommodations. I agreed immediately.

"One more thing," John said. "These Indians were killing oil prospectors just five years ago. You still want to go?"

This time my yes came a little slower. "Uh, well, uh, okay, yes. Let's do it!"

We flew into Quito, the capital of Ecuador. Then we took a smaller plane to a remote airstrip cut out of the rain forest near a small oil boom town called Argo Lago. It was little more than a collection of huts and stores that had sprung up overnight when oil had been discovered. The forest all around the town had been cut to supply lumber for the buildings. There must have been a lawless element, for there was also a town marshall who walked around with a pistol in his belt and a rifle in his hands. We caught a bus out of town and bounced along a dirt road for about an hour. We crossed two rivers, one on a hand-drawn ferry. The other was shallow enough for the bus to drive through.

Finally, we came to a stop on the banks of a large river. The Napo River is a tributary of the Amazon River. The river's current was quite swift, and the water was clear and deep. Our boat was waiting for us. It was what is called a **dugout**, made from a single tree trunk that had been hollowed out to allow for passengers and their freight. It was large enough to hold about 12 people and all their supplies. The front end was pointed, and the rear end was squared off. It was powered by an outboard motor. There were two boatmen. One sat up front and looked for sandbars and floating things in the water, like logs. The second one sat at the back and ran the motor.

As soon as we left the clearing, we left civilization behind. The rain forest crowded the banks of the river. Swallows and orioles flew overhead. Parrots and parakeets called loudly in the trees. After about an hour, the boatmen slowed down and turned into the bank. There was a dim path leading into the forest.

We all grabbed our packs, binoculars, and other gear and climbed out of the boat. The boatmen didn't wait around. They revved up the motor and headed back upstream. I looked back at the path. There was an Indian standing there. The first thing that I noticed was the old flintlock rifle that he held over his shoulders. He had on an old pair of shorts, and that was all. There was a hugh old scar that ran up his arm to his shoulder. It looked as if long ago, something had bitten a big chunk out of his arm, but the arm worked fine. He pointed down the trail. We loaded up our gear and began to follow along behind him.

The trees were very large with huge vines running up them. No sunlight reached the path, which was covered with dead leaves. Off the path, there were huge ferns and palms growing under the trees. Suddenly, a loud roar stopped us in our tracks. We all thought it was a jaguar, but the Indian pointed overhead. High on a fig tree, there was a group of large, black monkeys. We realized why they are called Howler Monkeys, as the large male roared at us again. We continued down the trail. The buzz of the insects and the calls of the tree frogs drowned out the chirps of the birds.

Finally, we came out on an **oxbow lake**. An oxbow is an old bend in a river that was cut off when the river changed course. It created an oxbow or U-shaped lake. Waiting for us there was another dugout. This one was smaller than the first, and it had a leak. Two Indians jumped out of the boat to help us load our packs.

An old Indian, still sitting in the boat, was fishing with just a piece of line and a hook. His bait was a little green worm. In the boat at his feet were three hand-sized fish. Fish that had teeth—piranha! To attract the fish, he would splash his hand on top of the water, maybe immitating a baby bird splashing in the water. He dropped the worm into the disturbed water, and soon he had another fish flopping in the boat.

We loaded up the boat and pushed off into the lake. The two older Indians picked up paddles and we headed across the lake. As we neared the far shore, we ran up on a sandbar, and we were stuck! There was a discussion among the Indians in their native dialect, of which we understood nothing. The young Indian kids, however, understood quite well, and they didn't seem very happy. Over the side they went, into the piranha-infested water. They pushed the dugout off of the sandbar, then quickly jumped back into the boat.

Finally, we reached a high bank on the lake's edge. Steps had been cut into the clay. We clambered up the hill to a clearing. In the middle of the clearing was a long thatched-roof cabin. It had a screen door and screened windows that ran the length of the building. Along the wall, there was a row of benches with straw mattresses. This would be our home for the next few days.

We soon established a daily routine. The soft calls of the Mottled Owl would stop as daylight approached. Often, we awoke to the roars of the Howler Monkeys or the raucous calls of parrots and macaws. Breakfast was usually eggs and bananas; lunch was rice and beans; dinner would be rice, beans, cooked bananas, and meat. The meat was sometimes Javaline, a pig-like animal; fish, maybe piranha; or Gibnut. The Gibnut was a jungle rodent the size of a rabbit that looked like a rat. It was all very tasty.

After an early breakfast everyday, we would hit the trails looking for birds. Finding more than 500 species was a possibility! As we birded along, we recognized many of the birds from the field guides we had been studying. One little nondescript flycatcher, however, puzzled us. We recorded its call and later found out that it was a newly-discovered species called the Orange-eyed Flycatcher.

Some days, the old, scarred Indian would lead us on our walks. He would point out interesting things such as Leaf-cutter Ants, an eight-inch Damsel Fly that looked like a helicopter when it flew, a colorful frog, and some huge black butterflies. Once, he showed us a big brown glob of dirt up in a tree. It was a Marching Ants nest. When he clapped his hands, each ant made a tiny munching sound. There were so many of them that it actually sounded like an army marching in step through the woods. One of the birders in our group had a sore throat. The old Indian gave her a piece of leaf to chew, and the sore throat went away. The rain forest really is the Indians' drugstore.

One day, the Indian boys peeled some bark off of a tree limb. They took some charcoal from the fire pit and proudly wrote their names on the white wood. They had gone to a one-room school house on the Napo River. It was about a two-hour trip to school by dugout and hiking trails. They didn't go every day, but they seemed very proud of what they had learned.

Time flew by as we discovered all the wonderful new things the rain forest had to offer. Soon it was time to leave. One of the birdwatchers was going to give the boys some money as a tip for helping us. Our guide pointed out that they had no place to spend the money, so we left them some pencils and a notebook, instead.

Within a few hours, we had retraced our journey back down the trails, across the lake, up the river, and back to civilization. We had left the peaceful rain forest where life had changed very little in the last hundred years. We found ourselves back in a large city with skyscrapers, freeways, and too much traffic. We put our hands over our ears, trying to block out the noisy cars and trucks and trying to hold in the memories of the monkeys and birds calling in the forest.

Conserving the Tropical Rain Forests

We hear a lot about the destruction of the tropical rain forests. Different experts come up with a wide range of figures, but they all agree that the problem is critical. Approximately 30,000 square miles of rain forest are being destroyed each year. At that rate, the rain forest will be gone in 50 years.

Since the 1960s, 75 percent of the forests of Africa's Ivory Coast have been cut down. India and Sri Lanka have had almost all of their primary rain forests logged. Overall, it is thought that half of the earth's rain forests are now gone. What is causing this destruction?

Our overpopulation is listed as the cause, and this might be true in some areas of Africa and Asia, but usually countries with the lowest populations have the largest rain forests, and they are still being destroyed. There must be some other reasons, then, other than overpopulation.

The biggest threats to the forests currently come from timber, mining, and other industrial activities. Agriculture, especially the slashing and burning of large areas to produce cattle pastures has destroyed a lot of rain forests. Hundreds of species of plants and animals have become extinct even before scientists have had a good chance to study them.

So why should we care? After all, the rain forest is a long way away. This destruction does not affect us ... or does it? The burning of rain forests adds carbon dioxide to the atmosphere. This can contribute to global warming. Destruction of trees can contribute to long-term droughts. Lack of soil cover promotes flooding. Already, thousands of medicines have been derived from tropical plants. There may be a plant that holds the key to curing cancer, but it may be destroyed forever before we can find it.

The problem is severe, and often it seems impossible, but there is hope. It was discovered that some fast food outlets were buying beef from land that used to be rain forest. Environmental groups put on the pressure, and most of those outlets now purchase **domestic beef**. (That means beef that has been grown in the United States.)

The World Bank was providing money to build roads in the rain forests. The roads allowed easy access for loggers and settlers who had no idea how to protect rain forest soils. Most of these projects were disasters, producing only short-term gains. The World Bank now looks at the environmental aspects of projects before funding them.

Conservation organizations have come up with another plan. Many poor countries have huge foreign debts. Conservation groups agree to purchase the debt from a bank at a reduced rate. The country agrees to protect tropical rain forest lands. This program is working in Ecuador, the Philippines, Belize, Costa Rica, and other countries.

When the rain forest is removed, tropical rains soon wash away top soil and leach nutrients out of the planting zone. Replanting a rain forest with all of its different species is impossible. Work is being done to see if damaged rain forests can be restored. We know that over very long periods of time, rain forests can reclaim some areas. We see the rain forests growing on areas cleared by the Mayas 1,000 years ago.

So what can you do to help protect the rain forest? You have already started by learning more about them. You can help by supporting conservation organizations that are working on the problem. Write to these groups and ask for information. While you are in a writing mood, write to your senators and representatives and ask them to support programs that protect the forests.

Maybe your class can contact students your age in tropical countries and see how they feel about the rain forest. The Student Letter Exchange at 630 Third Avenue in New York, New York, 10017, might help you get in contact with such students.

By recycling paper and other products, we can help take the pressure off some of our resources. Using less paper means fewer trees need to be cut.

We can set a good example by increasing the number of trees in our own environments, like our backyards. Talk your mom and dad into planting more trees in the yard. See if there is room on the school grounds for more trees, as well.

Finally, we need more scientists to study the tropical rain forest. Wouldn't you like to spend some time in one of the most diverse forests in the world? There may be an important species of plant or animal waiting for you to discover it!

Name ______________________________ Date ____________

For the Student:

1. How many square miles of rain forest are being destroyed each year?

2. What countries have lost a majority of their rain forests?

3. What are the biggest threats to rain forests?

4. How does destruction of the rain forest affect us?

5. How could building roads hurt the rain forests?

6. Can a rain forest be replanted?

7. If a rain forest areas is cleared, does it always remain clear? How do we know?

8. List some things you can do to help protect rain forests.

Organizations Working to Protect Rain Forests

The following is a list of some organizations that are working to protect the tropical rain forest and may have more information for you.

The Forest Research Institute of Malaysia
Kepang 52109
Kuala Lumpur, Selangor, Malaysia

African Wildlife Foundation
1717 Massachusetts Ave.
Washington, D.C. 20036

Wildlife Conservation International
New York Zoological Society
Bronx, NY 10460

Program for Belize
P.O. Box 385X
Vineyard Haven, MA 12568

Pronatura
Merida, Mexico

Conservation International
1015 18th St., NW
Washington, D.C. 20036

Global Tomorrow Coalition
1325 G St., NW
Suite 915
Washington, D.C. 20005

International Union for the Conservation of Nature and Natural Resources
Avenue Mont Blanc
1196 Gland
Switzerland

National Audobon Society
645 Pennsylvania Ave.
Washington, D.C. 20003

National Wildlife Foundation
1400 16th St., NW
Washington, D.C. 20036-2266

The Nature Conservancy
1815 North Lynn St.
Arlington, VA 22209

Rain Forest Action Network
300 Broadway
Suite 28
San Francisco, CA 94133

Smithsonian Tropical Research Institute
APO
Miami, FL 34002-0011

World Wildlife Fund/Conservation Foundation
1250 24th St., NW
Washington, D.C. 20037

Missouri Botanical Gardens
P.O. Box 299
St. Louis, MO 63166

Name ______________________________ Date ______________________

Activity: Let's Have a Tropical Party

After studying the tropical rain forest, plan a class party. Bring in only products from the tropical rain forest.

For example:

trail mix (cashews, peanuts, banana chips, chocolate chips, dried pineapple)

coconuts

orange juice and Quinine wat

lemon-lime soda

avocado dip

cinnamon toast

grapefruit

hot chocolate

Check the list of tropical foods on page 43 and make a list of your favorites.

____

____

____

____

____

____

____

____

____

____

____

____

____

____

Answer Keys

What Is a Tropical Rain Forest (page 3)
1. A tropical rain forest has warm temperatures all year, more than 60 inches of rain per year, and very high humidity.
2. No, they are found in a narrow band around the earth's equator.
3. The Sun's rays hit the earth at right angles at the equator.
4. No, those states are not located in the tropics.
5. The plants of the rain forest release huge amounts of water vapor, and this can affect global weather patterns.

6. E
7. D
8. C
9. A
10. B

Off to the Tropical Rain Forest (page 5)
1. Brazil, Zaire, and Indonesia have the most tropical rain forest.
2. Climatic changes and people cutting and burning trees have reduced rain forests.
3. No, the United States is not in the tropics.
4. The tropical seasonal forest has a dry season and a wet season.
5. Moisture comes from mists and clouds in the mountains.

Regions of Tropical Rain Forest (page 7)
1. Asia has the greatest diversity of plant and animal species because populations are isolated on remote islands, and they produce new species.
2. No, they are not found in the South American rain forests.
3. Mountains and water, such as oceans isolating islands.

4. C
5. B
6. C
7. D
8. A
9. B
10. A
11. C
12. B
13. A
14. C
15. A

Travel in the Rain Forest (page 8)

1. 1	18. 1
2. 3	19. 1
3. 1	20. 2
4. 1	21. 2
5. 3	22. 1
6. 2	23. 2
7. 3	24. 3
8. 2	25. 1
9. 2	26. 3
10. 1	27. 1
11. 1	28. 2
12. 3	29. 1
13. 1	30. 1
14. 2	31. 2
15. 1	32. 3
16. 1	33. 3
17. 2	34. 1

Life Layers (page 11)
1. The four levels are the emergent layer, the canopy, the understory or shrub layer, and the forest floor.
2. The emergent layer receives the most sunlight. The forest floor receives the least sunlight.
3. Very little sunlight reaches the forest floor, so plants that depend on the sun to provide food for survival do not grow well there.
4. They receive more sunlight up on the tree branches than on the forest floor.
5. It is a point on the end of a leaf that allows the water to run off. It keeps the leaves dry so molds and lichens do not grow on them.
6. Buttresses are thick, vine-like structures on emergent trees that help support these wind-exposed trees.
7. The canopy level has the most stable condition.
8. There are usually only one or two per acre.
9. Ferns, herbs, mosses, and fungi do well on the forest floor.
10. They have large leaves to capture as much sunlight as possible.

Biodiversity (page 13)
1. The number of living organisms in a habitat or in a geographic area is known as its biodiversity.
2. There are more birds.
3. Look for something very small in an unexplored area, like the rain forest or the ocean.
4. A ten-acre marshy wetland has a greater biodiversity.
5. Answers will vary.
6. The tropical rain forests have the greatest biodiversity.
7. It may affect other species.

8. These organisms are small and hard to find, and they have not been studied much until recently.

Plants of the Tropical Rain Forest (page 15)
1. F
2. T
3. F
4. F
5. T
6. F
7. T
8. T
9. T
10. F

11. A niche is a particular place that meets a plant's needs for survival.
12. They have an extra layer of bark that can slough off when the weight of the epiphytes gets to be too much.
13. Aerial roots are roots of epiphytes that absorb water from the moisture-laden air.
14. Seedlings grow rapidly to fill the gap, and other seeds germinate quickly and grow rapidly to fill the gap.

Plant Defenses (page 17)
1. It helps the leaf stay drier and makes it harder for mold and fungi to grow on the leaf.
2. The chemicals discourage leaf-eating animals.
3. The large leaves allow plants to survive in the low light of the forest floor.
4. The cup holds water for the plant to use. It also is a place for insects and frogs to lay their eggs. The residue left by the insects and frogs provides the plant with nutrients.
5a. that which is left over after part is taken away
b. a nutritious ingredient or substance in a food
c. plants that are parasites on living organisms or feed upon dead organic material
6. They eat the sugary substance the plant produces, and they bite anything that moves into the area to protect the source of the sugar.
7. Their mouths get gummed up.
8. It prevents mold and other fungi from being able to "hold on" to the leaves.

Animals of the Tropical Rain Forest (page 19)
1. It grows a mat of green algae on its fur.
2. Collared and White-lipped Peccarys travel in small herds or family groups.
3. Very little wind reaches the forest floor.
4. The jaguar is the largest cat of the tropical rain forest.
5. Visibility is reduced due to low light. There are many tree trunks, branches, and other types of foliage to try to see through. Animals may be camouflaged. There are lots of different kinds of animals, but individuals of a particular species may be scattered over a large area.
6. Jaguars have mottled spots.
7. The tapir, White-tailed Deer, and Brocket Deer are found on the forest floor.
8. The Black Howler Moneky, toucan, and Harpy Eagle are found in the upper canopy.
9. The Coati and Tree Boa may be found in several different levels of the rain forest.
10. They are captured and fitted with a small radio transmitter.

An Abundance of Birds (page 21)
1. About 700 species are found in North America.
2. About 2,900 species are found in South America.
3. More than 300 species were reported in one day.
4. Flocks of parrots and macaws gather at mineral clay banks along some rivers.
5. They are there to eat the minerals that help neutralize the poisons found in some fruits they eat.
6. You are more likely to hear them, because they are shy and secretive.
7. The bird was found next to a busy, four-lane highway in Brazil.
8. No

Foraging Flocks (page 25)
1. Only a couple of individuals from each species are found in the flock.
2. The flock may contain 20, 30, or even 60 species.
3. It is a noisy and alert bird that keeps a look out for predators and warns the others with its loud call.
4. It looks for insects under dead leaves.
5. The Stony-billed Woodcreeper knocks bark off trees.
6. The Red-legged Honeycreeper searches for nectar.
7. An ornithologist is a scientist who studies birds.
8. They travel together for protection, so one can warn the others of danger. They also may get more food, since one bird may stir up insects that another bird can eat.

Ants and Plants (page 27)
1. F
2. F
3. T
4. T
5. T
6. T

7. F
8. F
9. T
10. F

Word Search

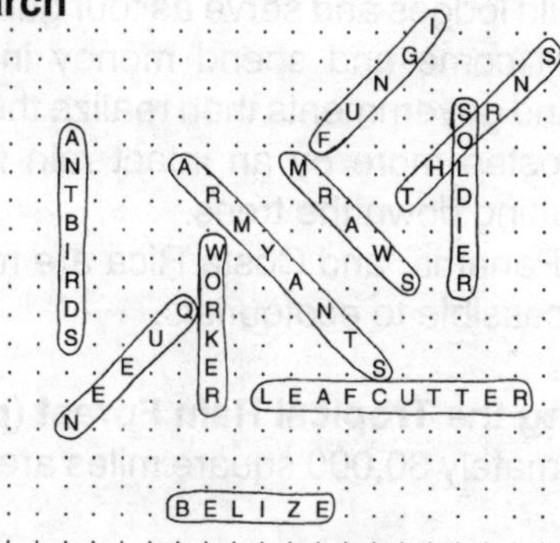

Hot and Humid, Dark and Dangerous (page 29)
1. St. Louis gets hotter.
2. Small parasites are more dangerous because they can spread disease without being seen.
3. The poison comes from the skin of poison tree frogs.
4. A caiman is a tropical member of the alligator family.
5. No.
6. A person must come in contact with the leaves or sap of the tree.
7. Jaguars, ocelots, and pumas are found in the tropical rain forest.
8. Yellow fever, tyfus, malaria, and hepititis are more prevalent in the tropics.
9. You can get shots and take precautions.
10. No.

People of the Rain Forest (page 32)
1. Amazonian Indians, Mayas, Pygmies, Penau, and Lua live in the rain forests. (any three)
2. They used bows and arrows, blow guns, spears, and traps.
3. All the trees are cut in an area and left to dry. Then they are burned and a crop is planted. After a few years, the field is abandoned.
4. After about 20 years, the area may be cleared again and turned into a field.
5. There is reduced forest habitat and diseases have been introduced.
6. Eighty-seven tribes have died out.
7. They have to wait for their crops to mature, and they don't have to follow the animals.
8. They trade Brazil nuts, coffee, and bananas.
9. They get meat from the animals of the forest, and they gather fruits and plants for the rest of their food. They use forest resources to build their huts, and they use forest plants to treat their illnesses.

Lost Cities (page 35)
1. It was named after the female warriors of Greek mythology because Francisco De Orellana reported the existence of fierce female warriors while sailing up the river.
2. He penetrated far into the interior, but he and all of his men disappeared without any trace in 1924.
3. They call it the elephant killer.
4. They are 20 to 25 feet long.
5. The Mayas built the great cities of Belize.
6. There were symbols telling stories of the king's personal accomplishments and the city-state's accomplishments.
7. They did not use wheels or draft animals.
8. There are 200,000 today. There were 2,000,000 in the time of the Mayas.

Great Rain Forest Drains and the Amazing Amazon (page 39)
1. The Amazon, Congo, and the Nile drain rain forests.
2. It's path, which crosses the equator, evens out its dry and wet seasons.
3. The Amazon River discharges more water.
4. They have a lot of water moving very quickly, so the silt doesn't slow down until it is far out into the ocean.
5. No, animals find more food, plants let water distribute seeds, and silt adds to soil fertility.
6. At times one-fifth of the world's river water is flowing in the Amazon.
7. At least 60 inches of rainfall a year occurs in a rain forest.
8. It is a large island at the mouth of the Amazon River.

Cycles and the Rain Forest (page 42)
1. Water is found in solid, liquid, and gaseous forms. Examples are ice, rain, and water vapor.
2. It can melt ice and snow, especially the polar ice caps.
3. Excessive carbon is released through burning fossil fuels and slash-and-burn agriculture.
4. It is expected to increase four to 11 degrees over the next 100 years.
5. It could cause changes in weather patterns and crop failures.
6.

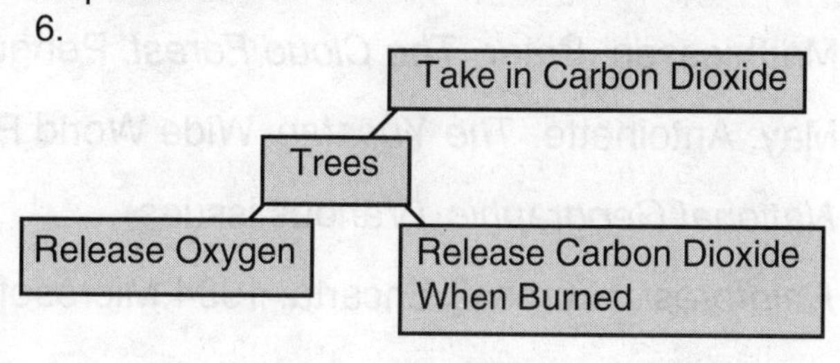

Our Most Important Forest (page 45)

1. More than half of the earth's species are located in the tropics.
2. The rain forest has been developing for 30 million years.
3. An estuary is an area where fresh water meets salt water, often found where rivers meet the sea.
4. It keeps silt from polluting the estuary.
5. The plants of the forest take up rain water through their roots, and it later evaporates from their leaves.
6. The birds won't have a place to spend the winter, and their populations will decline.
7. There aren't as many birds eating large numbers of insect pests.
8. Their homes and ways of life are destroyed.

The Bushdoctor Word Search (page 48)

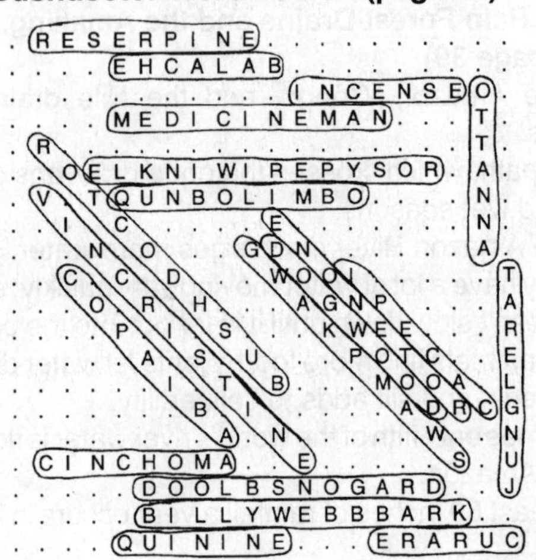

Ecotourism (page 50)

1. A tourist is someone who travels for pleasure.
2. Ecotourism is travel for pleasure to see some aspect of the modern natural world.
3. Answers will vary.
4. They build lodges and serve as tour guides hoping tourists will come and spend money in the area. Villagers and governments then realize the economy will be boosted more by an intact rain forest than through cutting down the trees.
5. Belize, Panama, and Costa Rica are making rain forests accessible to ecotourists.

Conserving the Tropical Rain Forest (page 56)

1. Approximately 30,000 square miles are destroyed each year.
2. Africa's Ivory Coast, India, and Sri Lanka have lost most of their rain forests.
3. Timber, mining, and other industrial activities are the biggest threats.
4. It adds carbon dioxide to the atmosphere, contributes to global warming, causes droughts and/or flooding, and destroys plants necessary for medicines.
5. The roads allow easy access for loggers and settlers who have no idea how to protect rain forest soils.
6. No
7. No, areas cleared by the Mayas 1,000 years ago are rain forest today.
8. Answers will vary.

Bibliography

Braws, Judy. *Nature Scope: Rainforests, Tropical Treasures.* National Wildlife Federation: Washington, D.C.

deSchauensee, R. Meger and Phelps, W. Jr. *Birds of Venezuela.* Princeton University Press, 1978. Princeton, NJ.

Forbath, Peter. *The River Congo.* Houghton Mifflin Co.: Boston, 1997.

Harwick, R. and Lyon, J. *The Belizian Rainforest.* Orangutan Press: Cray Mills, WI.

Matthiessen, Peter. *The Cloud Forest.* Penguin Books: New York, NY.

May, Antoinette. *The Yucatan.* Wide World Publishing: Tetra, 1993.

National Geographic. (Various issues)

Rainforest. Microsoft Encarta. 1994 Microsoft Corp.

Tropical Rainforests. R and E Online, Inc. 1996.